THE ELEMENTS

Nitrogen

John Farndon

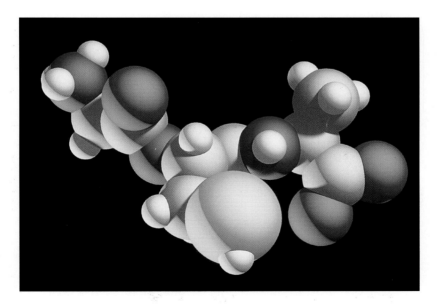

BENCHMARK BOOKS

MARSHALL CAVENDISH

NEW YORK

Benchmark Books
Marshall Cavendish Corporation
99 White Plains Road
Tarrytown, New York 10591-9001

Library of Congress Cataloging-in-Publication Data
Farndon, John.
Nitrogen / John Farndon.
p.   cm. — (The elements)
Includes index.
Summary: Discusses the origin, discovery, special characteristics, and use of nitrogen in
such products as explosives and fertilizers.
ISBN 0-7614-0877-0
1. Nitrogen—Juvenile literature. [1. Nitrogen.] I. Title. II. Series: Elements
(Benchmark Books)
QD181.N1F37 1999
546'.711—dc21   97-37945 CIP AC

Printed in Hong Kong

Picture credits
Corbis (UK) Ltd: 4, 6, 11, 19, 21, 30.
Science Photo Library: 7, 8, 9, 10, 13, 14, 15, 16, 17, 18, 20, 22, 23, 24, 25, 26, 27.

Series created by Brown Packaging Partworks
Designed by wda

Contents

Most of the air inside and around this balloon is nitrogen.

What is nitrogen?

Nitrogen is a remarkable substance. It surrounds us, yet we cannot see it, feel it, taste it, or smell it. Nitrogen does not even do very much—scientists say it is inert, because it does not normally react with other substances. But without nitrogen, life would be impossible.

Nitrogen compounds are a vital part of every living thing, since they are one of the main ingredients of the proteins from which cells are made. Every breath you take is mainly nitrogen, since nitrogen makes up almost four-fifths of air. The other main constituent is oxygen, and there are also small amounts of carbon dioxide and water vapor.

4

The nitrogen atom

Atoms are the building blocks of elements. They are far too small to see without a powerful microscope—the period at the end of this sentence would cover 250,000 million atoms.

At the center of each atom is a nucleus. This contains tiny particles called protons, which have a positive electrical charge. Nitrogen has an atomic number of seven, which means it has seven protons. The nucleus also contains neutrons, which have no charge. The neutrons and protons give nitrogen an atomic mass of 14.

Around the nucleus are even smaller negatively charged particles called electrons. The number of electrons is the same as the number of protons, so the nitrogen atom contains seven electrons.

ATOMS AT WORK

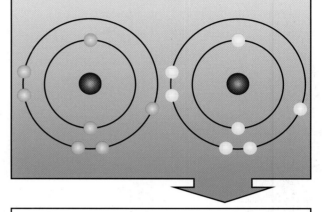

The seven electrons of a nitrogen atom are arranged in two shells: an inner shell with two electrons and an outer shell with five electrons. But atoms are only stable if they have eight electrons in their outer shell.

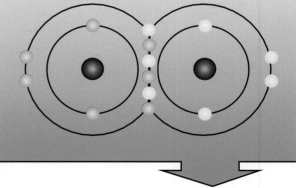

When one nitrogen atom meets another, each shares three of its electrons with the other, so that both atoms end up with eight electrons. The two atoms are locked together to form a stable molecule. This stable molecule makes nitrogen gas unreactive.

This link-up, which involves sharing electrons, is called a covalent bond. The bond in a nitrogen molecule is a triple bond, because it has three pairs of electrons. The symbol for nitrogen is N, and the molecule can be written like this:

$$N \equiv N$$

NITROGEN ATOM

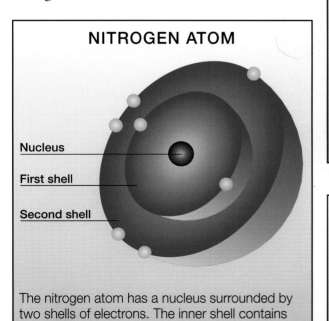

Nucleus

First shell

Second shell

The nitrogen atom has a nucleus surrounded by two shells of electrons. The inner shell contains two electrons. The outer shell has five.

Special characteristics

In nearly all circumstances on Earth, nitrogen is a gas. It does not condense (turn to liquid) until the temperature plummets to –321°F (–196°C). It does not freeze (turn to solid) until the temperature drops even further to –346°F (–210°C), which is far colder than anywhere ever found naturally on Earth. Pluto, the most distant planet in our solar system, is so far from the Sun that the daytime temperature on the planet's surface is about –355°F (–215°C). This chilly planet has an icy cover of frozen nitrogen.

Quick freeze

Liquid nitrogen is so cold that it can freeze things instantly to a degree never normally possible. This property makes it important in cryogenics. This is a field of science that works with very low temperatures. The normal freezing process takes so long that substances are damaged by the growth of ice crystals. But dipping them in liquid nitrogen freezes them so quickly that they suffer little damage.

Liquid nitrogen is also used in the food industry. Foods such as cheesecakes and soft fruits such as raspberries are often quick frozen with liquid nitrogen so they

DID YOU KNOW?

MAKING LIQUID NITROGEN

Nitrogen from the air can be turned into a liquid by a process called liquefaction. First carbon dioxide and water are removed from the air. Next the air is compressed and then allowed to expand. These steps are repeated over and over again. Every time the air expands, it cools down. With each expansion, the temperature drops a little further. Eventually, the air becomes so cold that it turns into a liquid. The nitrogen is drawn off and stored in strong steel cylinders.

This scientist is pulling a container out of a vat of liquid nitrogen. He is wearing gloves and a mask to protect himself from the extreme cold. The white smoke is liquid nitrogen turning into gas.

This picture shows the icy planet Pluto seen from its moon, Charon. The planet is covered with solid nitrogen.

can be stored in refrigerators. The foods are sent through a freezer on a conveyor belt. They are first cooled by nitrogen gas and then sprayed with liquid nitrogen.

Liquid nitrogen is used in hospitals to store blood until it is needed for operations. Doctors also use liquid nitrogen to remove tattoos, warts, birthmarks, and skin cancers.

Some people vainly hope that when they die their whole bodies may be preserved by freezing with liquid nitrogen until progress in science finds a means of bringing them back to life. However, it will be many, many years—if ever—before scientists can successfully freeze and unfreeze something as large and complicated as a human body.

Some of the nitrogen in the universe is found in nebulae, massive clouds of gas and dust in space.

Where nitrogen is found

Nitrogen is the sixth most abundant substance in the universe. It is found in stars, in nebulae (clouds of gas between stars), in the Sun, and in meteorites.

Most of the nitrogen in and around Earth is found in the atmosphere. There is very little nitrogen in Earth's crust (the outer layer) or in its hot interior. But some nitrogen is found in the clouds of gas that erupt from volcanoes and in mineral water springs. A mineral called sodium nitrate, also known as Chile saltpeter, is mined in quarries and contains nitrogen attached to oxygen and sodium metal.

All plants and animals have nitrogen in their bodily cells. The amount of nitrogen inside living things is small in terms of the universe, but it is essential to life.

SEE FOR YOURSELF

You can see for yourself the nitrogen content of the atmosphere with this simple experiment.

● Set up a candle inside a dish full of water and light the candle. (Important: do not light the candle without an adult's help.)

● Allow the candle to burn for a few minutes, then lower a glass jar over the candle at an angle, permitting any air bubbles to escape.

● Note the level of water in the jar. As the candle burns down, you will see the water level rising as the flame consumes the oxygen in the air.

● When all the oxygen is used up, the candle will sputter out. The gas that remains inside the jar is nitrogen. The water level will now be about a fifth of the way up the jar, showing that the air is made up of about one-fifth oxygen and four-fifths nitrogen.

How nitrogen was discovered

Two scientists discovered nitrogen quite separately in the early 1770s. They were Scottish physicist Daniel Rutherford (1749–1819) and Swedish chemist Carl Scheele (1742–1786). Rutherford named his discovery "noxious air," because animals were unable to breathe in it. Scheele called it "foul air."

But it was the great French scientist Antoine Lavoisier (1743–1794) who realized that air is essentially a mixture of two gases, oxygen and nitrogen. When Lavoisier burned the metal mercury in a closed jar, he found that a fifth of the air combined with the mercury to form a red powder, mercuric oxide. The rest of the air stayed a gas, no matter what he did.

DID YOU KNOW?

WHAT'S IN THE AIR

A few years after Lavoisier discovered that air was mainly a mixture of two gases, oxygen and nitrogen, British chemist Henry Cavendish (1731–1810) suggested that air might also contain small amounts of another gas, even less reactive than nitrogen. A century later, this extra-lifeless gas was detected and named argon, from the Greek for "inert." Argon makes up less than 1 percent of the atmosphere while oxygen makes up 21 percent and nitrogen 78 percent.

Later, air was found to contain minute traces of four other inert gases—neon (from the Greek for "new"), krypton (from the Greek for "hidden"), xenon (from the Greek for "stranger"), and helium (from the Greek for "sun").

In recent years, traces of still more gases have been found—nitrous oxide, methane, carbon dioxide, and carbon monoxide. Most of these gases come from automobile exhausts and from the burning of fossil fuels—coal, gas, and oil.

A candle would not burn in this remaining gas, and mice died in it.

Lavoisier decided that air is made of two gases. One, which he called oxygen, was the gas that burned with the mercury. The other he called azote from the Greek for "no life." It later came to be called nitrogen, because it can be generated from niter, the common name for sodium nitrate.

Antoine Lavoisier working in his laboratory.

Lightning flashes heat the air and provide enough energy to break the molecules of nitrogen apart.

How nitrogen reacts

Atoms of nitrogen are so tightly bound together in molecules that they will not normally react with anything else. However, under certain circumstances, nitrogen can be made to react.

Reactions with oxygen

Nitrogen atoms react with other substances when the bonds that hold them together as molecules are broken. This normally takes a huge amount of energy, such as exists in a lightning flash.

LIGHTNING FACTS

● Lightning strikes somewhere on Earth 100 times a second.

● Lightning usually happens when warm, moist air rises from the ground to form towering thunderclouds.

● The nitrogen dioxide that forms after a lightning strike dissolves in rainwater to form 250,000 tons of acid every day.

● There are three types of lightning. These are ball lightning, forked lightning, and sheet lightning.

The dirty brown haze that hangs over cities such as Los Angeles is nitrogen dioxide gas.

When lightning flashes through the sky, it heats the air, releasing so much energy that the nitrogen molecules in the air are split in two. The nitrogen atoms are now free to react with oxygen in the air. They combine with oxygen atoms to form the compound nitrogen monoxide. This is a colorless gas in which each molecule consists of one nitrogen atom and one oxygen atom.

Nitrogen and oxygen can react together to form a range of compounds. When nitrogen monoxide cools down, for example, it may react with oxygen in the

ATOMS AT WORK

Nitrogen dioxide is a brown gas in which the molecules are made from one nitrogen atom and two oxygen atoms. It normally forms at high temperatures.

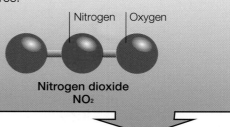

Nitrogen dioxide
NO₂

When nitrogen dioxide cools down, pairs of molecules may link together to form a pale yellow liquid called dinitrogen tetraoxide. Scientists call dinitrogen tetraoxide a dimer of nitrogen dioxide; a dimer is two identical molecules linked together.

The chemical reaction that takes place in the dimerization of nitrogen dioxide is written like this:

2NO₂ → N₂O₄

This tells us that two molecules of nitrogen dioxide combine to give one molecule of

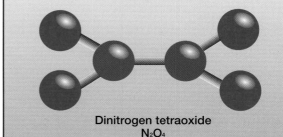

Dinitrogen tetraoxide
N₂O₄

THE HABER PROCESS

At the beginning of the 20th century, nitrogen compounds were already widely used as fertilizers and for explosives. But most of these compounds came from sodium nitrate deposits in Chile, which were rapidly running out. So chemists began to look for another source of nitrogen compounds. The solution, discovered by a German chemist named Fritz Haber (1868–1934), was to make ammonia by combining nitrogen from the air with hydrogen in what later came to be called the Haber process.

In this process, nitrogen and hydrogen are mixed in the proportions 1:3. The two gases combine only when they are compressed to over 200 times normal atmospheric pressure and they are passed over a catalyst of iron powder. (A catalyst makes a reaction occur faster.) To make the reaction go reasonably quickly, the temperature is usually boosted to 932°F (500°C).

Reactions with hydrogen

Nitrogen can be made to react with the gas hydrogen to form a pungent gas called ammonia. This is produced in factories by the Haber process, which uses very high temperatures and pressures to make the reaction occur. Bacteria manage to do the same under normal conditions by a process called nitrogen fixation (see page 15). Ammonia is used to make fertilizers, explosives, cleaning fluids, rocket fuels, and various drugs.

Just as the nitrogen oxides lead nitrogen into a range of reactions with other chemicals, so too does ammonia. Ammonia burns in oxygen to give nitrogen and water. It reacts with acids to form compounds called ammonium salts. Some of these salts are extremely useful. Ammonium chloride, which is formed when ammonia reacts with hydrochloric acid, is used in batteries. The reaction between ammonia and sulphuric acid makes ammonium sulphate. This salt is a good fertilizer. It is added to soil to make plants grow.

Reactions with metals

When it is heated to temperatures above 390°F (198°C), nitrogen reacts with metals to form compounds called nitrides. These compounds are not very stable, and they readily fall apart again when they are added to water.

air to form another compound called nitrogen dioxide. The molecules have one nitrogen atom attached to two oxygen atoms. Nitrogen dioxide is a poisonous, brown, acidic, pungent gas.

Nitrogen dioxide, in turn, reacts with water to form nitric acid and nitrous acid. Acids are very reactive, and so nitrogen is led in this way into a wide range of other chemical reactions.

The launch of the space shuttle Columbia on June 25, 1992. The fuel that powers spacecraft into orbit is made from ammonia.

ATOMS AT WORK

In the Haber process, ammonia is made from nitrogen and hydrogen. Normally, these two gases exist as molecules made of two atoms held together by strong bonds.

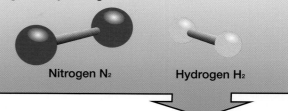

Nitrogen N₂   **Hydrogen H₂**

When they are mixed and compressed, the bonds break. The nitrogen and hydrogen molecules split apart into separate atoms.

When the mixture is passed over iron in a heated chamber, nitrogen atoms combine with hydrogen atoms to form an ammonia molecule.

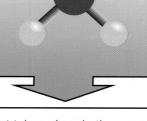

Ammonia NH₃

The chemical reaction that takes place in the Haber process is written like this:

N₂ + 3H₂ → 2NH₃

This tells us that one molecule of nitrogen combines with three molecules of hydrogen to give two molecules of ammonia.

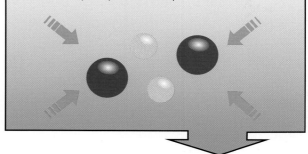

The grass in this field takes nitrogen compounds from the soil. The cows get their nitrogen by eating the grass.

The nitrogen cycle

Nitrogen atoms are constantly moving in a giant circle from the air, through the soil, into the bodies of plants and animals, and eventually back to the air. This whole process is called the nitrogen cycle.

All living things need nitrogen to develop and grow. Even though plants and animals are surrounded by huge quantities of nitrogen in the air, they cannot use any of it because the nitrogen atoms are far too firmly bound together in molecules. So plants must draw their nitrogen from nitrogen compounds dissolved in the soil, and animals get their nitrogen by eating plants or by eating other animals. But how does the nitrogen get into the soil?

Getting into the soil

A small quantity of the nitrogen found in the soil comes from nitrogen oxides made by lightning flashes. The nitrogen oxides dissolve in rainwater to form nitric acid, which washes into the ground.

The rest of the nitrogen comes from minute organisms called bacteria. Bacteria are the only living things capable of fixing (getting) nitrogen directly from the air.

The process is started by certain kinds of bacteria in the soil that can extract nitrogen from the air. Then other bacteria convert the nitrogen into nitrogen compounds called nitrates, in a process

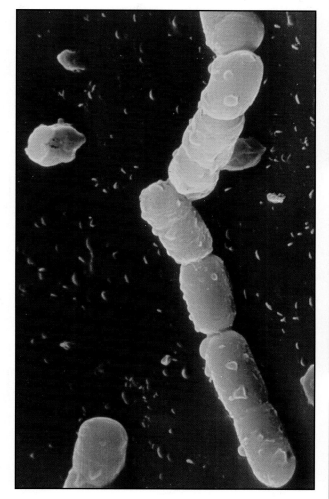

Tiny creatures called bacteria live in the soil. They take nitrogen from the air and use it for their food.

ATOMS AT WORK

Air contains nitrogen and oxygen. Both exist in the air as molecules. The atoms in molecules are held together by strong bonds. The atoms cannot react with anything else until these bonds are broken.

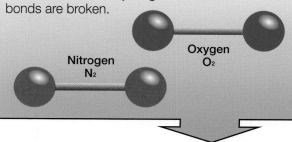

Nitrogen
N_2

Oxygen
O_2

When lightning flashes through the sky, it heats the air and gives it energy. This energy is enough to break the bonds in the nitrogen and oxygen molecules, pulling them apart.

The oxygen atoms and the nitrogen atoms are now free to join with other atoms and form new substances. The nitrogen atoms react with oxygen atoms to make nitrogen monoxide. Each molecule of nitrogen monoxide has one nitrogen atom and one oxygen atom.

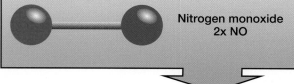

Nitrogen monoxide
2x NO

The chemical reaction that takes place during a lightning flash is written like this:

$N_2 + O_2 \rightarrow 2NO$

This tells us that one molecule of nitrogen combines with one molecule of oxygen to produce two molecules of nitrogen monoxide.

Legumes have tiny swellings on their roots. Bacteria living in the swellings take nitrogen from the soil.

called nitrification. Plants absorb the nitrates and turn them into more complex nitrogen compounds.

Legumes

A few plants can take nitrogen directly from nitrogen-fixing bacteria. These plants are called legumes. They include peas, beans, and clover. The bacteria live in little swellings or nodules on the plant's roots. This arrangement is good for the plants and the bacteria. The plants have a ready source of nitrogen, and the bacteria have a safe place to live.

Back to the air

Bacteria also play an important role in returning nitrogen to the air. Bacteria in the soil decompose animal droppings and animal and plant remains and produce ammonia. Nitrifying bacteria turn the ammonia into nitrates. Other bacteria, called denitrifying bacteria, convert some of the nitrates back into nitrogen gas, which is released into the air.

The cycle in balance

All these different steps form a massive cycle. The effect is that, over time, bacteria in the soil return almost the same amount of nitrogen to the air as other bacteria take from the air. This keeps the nitrogen content of Earth and its atmosphere in a perfect balance.

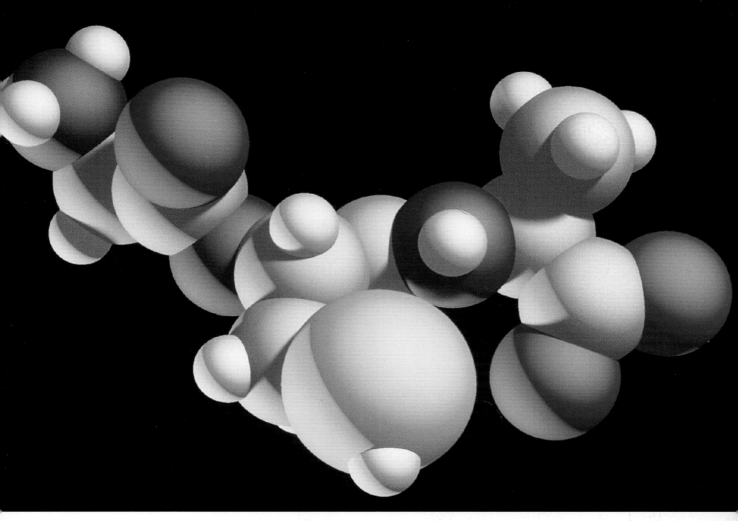

This is a computer model of three amino acids joined together. The nitrogen atoms are colored blue. The large white balls are carbon, the small white ones are hydrogen. Oxygen atoms are red and sulfur atoms are yellow.

Nitrogen in the body

Nitrogen compounds play a vital part in our bodies. Without nitrogen there would be no proteins, and without proteins our bodies would not exist. Some proteins are structural proteins, the building blocks from which the minute cells of our body are made. Others are enzymes that enable all the chemical reactions of the body to occur quickly and efficiently. Yet other proteins contain the instructions that tell our bodies how to make new cells.

Proteins are large, complicated molecules made from 20 or so different smaller molecules called amino acids. Amino acids are made from a combination of an amine and a carboxyl. An amine contains one nitrogen atom and two hydrogen atoms; a carboxyl is made of one carbon atom, two oxygen atoms, and one hydrogen atom.

Proteins and diet

Plants can make all their own amino acids, but our bodies cannot. Our bodies can only make 12 of the 20 amino acids. To stay healthy, we must get the other eight from our food. These eight are called the "essential" amino acids, and they come from the animal and plant proteins in the food we eat.

These proteins are broken down into amino acids, which are then absorbed and used to build new proteins. If we do not eat enough of the right proteins, our bodies are starved of the essential amino acids and become ill.

PROTEIN FACTS

● Good sources of protein include soybean flour (which is 40% protein), peanuts (28%), cheese (25%), meat (23%), fish (15%), eggs (12%), bread (8%), and rice (6%).

● Meat and fish contain all the essential amino acids but vegetables do not. To get the full range, people who do not eat meat must eat alternative foods, such as grain and beans.

● Children aged 9 to 12 need to eat around 2.8 ounces (80 grams) of protein every day.

● A single protein molecule is made from 500 or so amino acid molecules.

Our bodies cannot make their own proteins, so we have to get our protein from the food we eat. Good sources of protein include meat, fish, cheese, eggs, milk, nuts, beans, and cereals.

If divers return to the surface too quickly, the nitrogen in their blood forms tiny bubbles, which can be dangerous.

The "bends"

Nitrogen in the body can cause real problems for deep-sea divers. On a very deep or long dive, the extra pressure in the diver's lungs causes more nitrogen to dissolve in the blood than normal.

If the diver returns to the surface too quickly, this excess nitrogen can be dangerous. It forms tiny bubbles in the blood. The bubbles block the blood vessels, causing a painful and often fatal condition called the "bends."

The only way to deal with this problem is to put the diver into a decompression chamber, where the pressure is very high. The diver stays in the decompression chamber for hours while the pressure in the chamber is brought down gradually and the nitrogen bubbles slowly come out of solution and are breathed out.

Using nitrogen

One of the advantages of nitrogen is that it is so inert. So it is often used in place of oxygen, which is very reactive, in containers. Storage tanks containing ethanol (alcohol) are often filled with nitrogen rather than air because ethanol may burst into flames if it comes into contact with oxygen.

Perishable foods are often packaged with nitrogen, too. It is not air that makes potato chip bags look full but nitrogen. If the bags were filled with air, the oxygen in the air would react with the chips and make them go stale. The nitrogen in the bags also helps to stop the chips from being crushed.

Nitrogen compounds

Nitrogen compounds are also used in lots of ways. Modern fabrics get their bright colors from nitrogen compounds. These compounds are used to make dyes with strong orange, green, blue, and red colors.

Nitrous oxide, or dinitrogen oxide, contains one oxygen atom attached to two nitrogen atoms. It is a colorless gas with a sweet smell. It is sometimes called laughing gas, because breathing it makes people behave in a silly way.

English chemist Joseph Priestley (1733–1804) was the first person to make nitrous oxide. Later, another English scientist, Humphry Davy (1778–1829), studied the gas. When Davy breathed the nitrous oxide, he laughed and danced around the room.

The bright colors in these vats of dye are produced by nitrogen compounds.

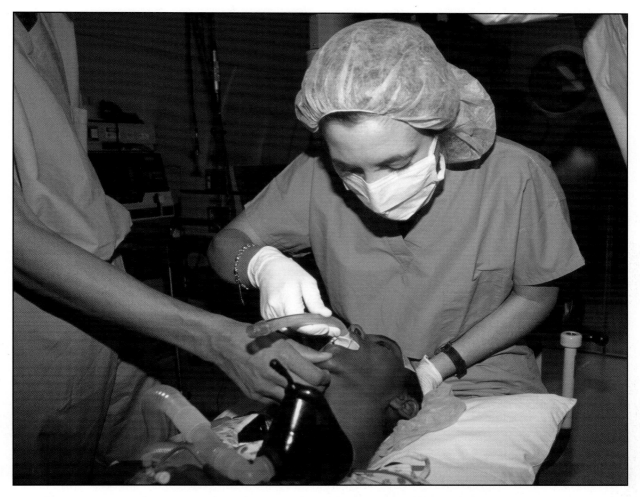

Breathing a mixture of nitrous oxide and oxygen during an operation keeps a patient unconscious.

The main use of nitrous oxide is as an anesthetic. Doctors and dentists use it to send patients to sleep during operations so that the patients feel no pain. Nitrous oxide works quickly, and its effects wear off rapidly, so it is easy to use.

Nitrous oxide is always given with oxygen because the body needs the oxygen to breathe—it cannot remove the oxygen atoms from the molecules of nitrous oxide.

DID YOU KNOW?

KEEPING VEHICLES SAFE

Nitrogen is inside the airbags that help to protect people from serious injury in vehicle crashes. The nitrogen is held in a compound called sodium azide, which has one sodium atom and three nitrogen atoms. When the airbag is activated, the sodium azide molecules very quickly fall apart. This releases nitrogen gas, which makes the airbag expand.

Explosives

One of the main uses of nitrogen is to make explosives. An explosive is like a chemical jack-in-a-box. But instead of a clown, an explosive has a store of energy inside it. When the explosive is set off, the energy bursts out in the form of a hot gas, which expands very quickly. All it takes to trigger the explosion is a tiny spark, a flame, or heat. Some explosives are set off if they are shaken or moved.

Fireworks on the fourth of July. The bright red colors are made by nitrogen compounds.

Explosives are used to blow up buildings. This building turns to dust and rubble in seconds.

One of the first explosives was gunpowder, which was invented about 1,000 years ago. The nitrogen in gunpowder is in a compound called potassium nitrate. When gunpowder is set off, it produces a cloud of oxygen and nitrogen gas. A small pile of black powder is left over.

Modern explosives

Nitrogen is still an important part of explosives today. These explosives include TNT and dynamite, which are used to blast rocks in mines and quarries and to demolish old or unsafe buildings.

ATOMS AT WORK

Gunpowder contains potassium nitrate. Each molecule of potassium nitrate is made of one potassium atom, one nitrogen atom, and three oxygen atoms.

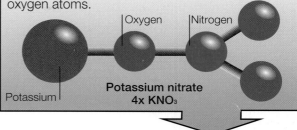

Oxygen | Nitrogen

Potassium

Potassium nitrate
4x KNO_3

When gunpowder is heated, the bonds in the potassium nitrate break apart. The atoms are free to join together to make other substances.

The reaction produces a massive cloud of gas, which is a mixture of nitrogen and oxygen. The gas is very hot and it expands quickly. It is this sudden burst of gas that causes the explosion. Most of the gunpowder disappears into the air as nitrogen and oxygen. All that is left is a tiny pile of potassium oxide, a black powder.

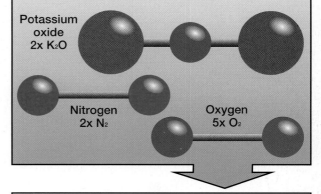

Potassium oxide
2x K_2O

Nitrogen
2x N_2

Oxygen
5x O_2

The reaction that takes place when gunpowder is set off is written like this:

$4KNO_3 \rightarrow 2K_2O + 2N_2 + 5O_2$

All chemical reactions have to balance. The number of nitrogen, oxygen, and potassium atoms is the same at the start and end of the reaction.

Fertilizers

Just as humans need to eat properly to stay healthy, so plants need certain nutrients to grow. Plants take nitrogen and other nutrients from the soil. The nitrogen usually returns to the soil when the plants die and decompose, or when their leaves fall to the ground and rot. The problem with farm crops is that they rarely put much back into the soil. Over the years, the soil becomes starved of nutrients and the crops fail to grow.

One way of replacing the lost nutrients is to rotate the crops. A farmer might plant a field one year with corn, wheat, and other crops that drain the soil of nitrogen. The next year, the field is planted with plants such as soybeans and peas that replace nitrogen.

But land is so expensive that farmers often want to grow the same crop year after year to maintain profits. In the past, farmers spread animal manure or guano (bird droppings) on the fields to help replace some of the lost nitrogen. But for most farmers these natural fertilizers are no longer adequate. They now use huge quantities of artificial fertilizers to boost crops to unheard-of levels.

Fertilizers add not only nitrogen but phosphorus and potassium too, all of which are vital for plant growth. Nowadays nitrogen fertilizers are made mainly from ammonia. Sometimes the ammonia is used as a liquid or pumped as a gas directly into the soil where it dissolves and helps plants grow. Usually, the ammonia is used to make nitrate compounds, which are spread on the soil as a powder.

Nitrate pollution

One of the problems with using all these nitrate fertilizers is they are washed down into the soil by rainwater. Unfortunately, they don't just disappear. They gradually

Farmers use fertilizers to improve their crops. Here, a farmer is injecting liquid ammonia into the soil.

The green film on the surface of this river is an algal bloom. It has been caused by a buildup of fertilizers.

seep through the ground over the years and end up in the world's rivers, lakes, and streams, polluting them.

The nitrates that build up in streams fertilize the plants in the water, making them grow much more quickly than usual. The plants use up so much oxygen that water creatures suffocate. This is called eutrophication. Sometimes, nitrate pollution can cause massive growths of algae, called algal blooms, which choke the life out of other wildlife such as fish by using up all the oxygen.

Nitrate pollution can also affect drinking water, since normal water purification does not eliminate it. It may not affect adults much, for the body can safely absorb some substances. But it can damage the kidneys of small babies.

NITRATE FACTS

● There are many different kinds of nitrates. They all dissolve easily in water and react with concentrated sulphuric acid to make nitric acid.

● Sodium nitrate is also called Chile saltpeter. As well as being used to make fertilizers and explosives, it is also added to foods to stop them from going bad.

● Potassium nitrate, or saltpeter, is the basis of gunpowder and is used to make fireworks.

● Silver nitrate is sensitive to the light. It is grains of silver nitrate on photographic film that record the picture. When the picture is developed, the exposed grains turn to silver.

● Strontium nitrate is used for making red distress flares.

Factories and mills produce nitrogen oxides, which are released into the atmosphere where they pollute the air.

Pollution

Nitrogen may be one of the main ingredients of the atmosphere but nitrogen compounds are some of the worst air pollutants. As human activities dominate our planet, so the atmosphere is becoming dirtied more and more with nitrogen compounds.

Noxious gases
Wherever oil, coal, and gas are burned—in power stations, automobiles, and homes, schools, and offices—oxides of nitrogen are released into the atmosphere. Nitrogen monoxide is made in the heat of an automobile engine just as it is by the searing heat of lightning. As nitrogen monoxide cools, it forms nitrogen dioxide.

This is one of the main ingredients of smog, the dirty, brown haze that hangs over many of our cities.

Nitrogen dioxide is harmful if it is breathed into the body, especially by young children and elderly people. And nitrogen oxides dissolve in rainwater and turn it into nitric acid.

Acid rain
All rain is slightly acidic, but human activities are making it more so. The three main culprits are: sulfuric acid from the sulfur dioxide released by burning oil, coal, and gas (65 percent); nitric acid from the nitrogen oxides produced in automobile exhausts (30 percent); and ozone, which forms in the reaction between nitrogen oxides and oxygen in the air (5 percent).

Acid rain kills fish, damages crops and trees, and destroys buildings. Millions of fish have died in the lakes and streams of the Adirondack Mountains in eastern New York, and trees have been destroyed in forests in Canada and northern Europe. Acid rain reacts with limestone and marble, which have been used as building materials for thousands of years. Among the buildings and monuments that have been damaged by acid rain are the Lincoln Memorial in Washington, D.C., and the Taj Mahal in India.

Many countries are taking steps to prevent acid rain. In the United States, for example, all new cars must be fitted with special converters that clean up waste gases. Power stations are being made to reduce the amount of sulfur dioxide and nitrogen oxides they send into the air. But more needs to be done to stop rain from becoming more and more acidic.

This forest is in the Czech Republic in central Europe. The trees have been killed by acid rain.

ATOMS AT WORK

In cars and power stations, nitrogen atoms and oxygen atoms combine to form nitrogen monoxide. When nitrogen monoxide cools, another oxygen atom joins on to make nitrogen dioxide.

Nitrogen | Oxygen

Nitrogen dioxide NO₂

The water molecules in rain are made from two hydrogen atoms and one oxygen atom. When nitrogen dioxide dissolves in rain, one hydrogen atom and one oxygen atom are added to make nitric acid.

Oxygen

Hydrogen

Water H₂O

The chemical reaction that takes place when nitrogen dioxide dissolves in rainwater is written like this: $2NO_2 + 2H_2O \rightarrow 2HNO_3 + H_2$

This tells us that two molecules of nitrogen dioxide combine with two molecules of water to give two molecules of nitric acid plus one of hydrogen.

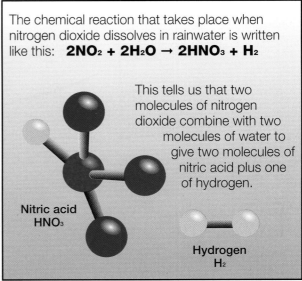

Nitric acid HNO₃

Hydrogen H₂

Periodic table

Everything in the universe is made from combinations of substances called elements. Elements are the building blocks of matter. They are made of tiny atoms, which are much too small to see.

The character of an atom depends on how many even tinier particles called protons there are in its center, or nucleus. An element's atomic number is the same as the number of protons.

Scientists have found around 110 different elements. About 90 elements occur naturally on Earth. The rest have been made in experiments.

All these elements are set out on a chart called the periodic table. This lists all the elements in order according to their atomic number.

The elements at the left of the table are metals. Those at the right are nonmetals. Between the metals and the nonmetals are the metalloids, which sometimes act like metals and sometimes like nonmetals.

- On the left of the table are the alkali metals. These elements have just one electron in their outer shells.

- Elements get more reactive as you go down a group.

- On the right of the periodic table are the noble gases. These elements have full outer shells.

- The number of electrons orbiting the nucleus increases down each group.

- Elements in the same group have the same number of electrons in their outer shells.

- The transition metals are in the middle of the table, between Groups II and III.

Group I

Group II

Transition metals

1 H Hydrogen 1								
3 Li Lithium 7	4 Be Beryllium 9							
11 Na Sodium 23	12 Mg Magnesium 24							
19 K Potassium 39	20 Ca Calcium 40	21 Sc Scandium 45	22 Ti Titanium 48	23 V Vanadium 51	24 Cr Chromium 52	25 Mn Manganese 55	26 Fe Iron 56	27 Co Cobalt 59
37 Rb Rubidium 85	38 Sr Strontium 88	39 Y Yttrium 89	40 Zr Zirconium 91	41 Nb Niobium 93	42 Mo Molybdenum 96	43 Tc Technetium (98)	44 Ru Ruthenium 101	45 Rh Rhodium 103
55 Cs Cesium 133	56 Ba Barium 137	71 Lu Lutetium 175	72 Hf Hafnium 179	73 Ta Tantalum 181	74 W Tungsten 184	75 Re Rhenium 186	76 Os Osmium 190	77 Ir Iridium 192
87 Fr Francium 223	88 Ra Radium 226	103 Lr Lawrencium (260)	104 Unq Unnilquadium (261)	105 Unp Unnilpentium (262)	106 Unh Unnilhexium (263)	107 Uns Unnilseptium (?)	108 Uno Unniloctium (?)	109 Une Unillenium (?)

Lanthanide elements

Actinide elements

57 La Lanthanum 39	58 Ce Cerium 140	59 Pr Praseodymium 141	60 Nd Neodymium 144	61 Pm Promethium (145)
89 Ac Actinium 227	90 Th Thorium 232	91 Pa Protactinium 231	92 U Uranium 238	93 Np Neptunium (237)

The horizontal rows are called periods. As you go across a period, the atomic number increases by one from each element to the next. The vertical columns are called groups. Elements get heavier as you go down a group. All the elements in a group have the same number of electrons in their outer shells. This means they react in similar ways.

The transition metals fall between Groups II and III. Their electron shells fill up in an unusual way. The lanthanide elements and the actinide elements are set apart from the main table to make it easier to read. All the lanthanide elements and the actinide elements are quite rare.

Nitrogen in the table

Nitrogen has an atomic number of seven, which tell us it has seven protons in its nucleus. This makes nitrogen one of the lightest elements. It is in Group V, which means it has five electrons in its outer shell. Nitrogen has a very low boiling point and freezing point and usually occurs as a gas.

Metals

Metalloids (semimetals)

Nonmetals

7	Atomic (proton) number
N	Symbol
Nitrogen	Name
14	Atomic mass

Group VIII

			Group III	Group IV	Group V	Group VI	Group VII	2 He Helium 4
			5 B Boron 11	6 C Carbon 12	7 N Nitrogen 14	8 O Oxygen 16	9 F Fluorine 19	10 Ne Neon 20
			13 Al Aluminum 27	14 Si Silicon 28	15 P Phosphorus 31	16 S Sulfur 32	17 Cl Chlorine 35	18 Ar Argon 40
28 Ni Nickel 59	29 Cu Copper 64	30 Zn Zinc 65	31 Ga Gallium 70	32 Ge Germanium 73	33 As Arsenic 75	34 Se Selenium 79	35 Br Bromine 80	36 Kr Krypton 84
46 Pd Palladium 106	47 Ag Silver 108	48 Cd Cadmium 112	49 In Indium 115	50 Sn Tin 119	51 Sb Antimony 122	52 Te Tellurium 128	53 I Iodine 127	54 Xe Xenon 131
78 Pt Platinum 195	79 Au Gold 197	80 Hg Mercury 201	81 Tl Thallium 204	82 Pb Lead 207	83 Bi Bismuth 209	84 Po Polonium (209)	85 At Astatine (210)	86 Rn Radon (222)

62 Sm Samarium 150	63 Eu Europium 152	64 Gd Gadolinium 157	65 Tb Terbium 159	66 Dy Dysprosium 163	67 Ho Holmium 165	68 Er Erbium 167	69 Tm Thulium 169	70 Yb Ytterbium 173
94 Pu Plutonium (244)	95 Am Americium (243)	96 Cm Curium (247)	97 Bk Berkelium (247)	98 Cf Californium (251)	99 Es Einsteinium (252)	100 Fm Fermium (257)	101 Md Mendelevium (258)	102 No Nobelium (259)

Chemical reactions

Chemical reactions are going on all the time—candles burn, nails rust, gasoline ignites in automobile engines, food is digested. Some reactions involve just two substances; others many more. But whenever a reaction takes place, at least one substance is changed.

In a chemical reaction, the atoms do not change. An iron atom remains an iron atom; an oxygen atom remains an oxygen atom. But they join together in new combinations to form new molecules.

Writing an equation

Chemical reactions can be described by writing down the atoms and molecules before and the atoms and molecules after. Since the atoms stay the same, the number of atoms before will be the same as the number of atoms after; only the molecules change. Chemists write the reaction as a chemical equation.

Equations are a quick and easy way of showing what happens in a chemical reaction. They use chemical symbols, rather than words, so scientists all over the world can understand them.

Making it balance

When the number of each atoms on both sides of the equation are equal, the equation is said to be balanced. If they are not equal, something must be wrong. So the chemist looks at the equation again and adjusts the number of atoms involved until the equation balances.

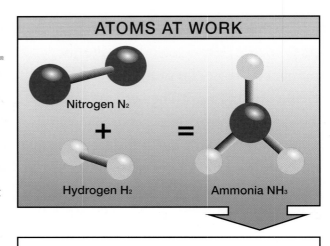

ATOMS AT WORK

Nitrogen N$_2$

+

Hydrogen H$_2$

=

Ammonia NH$_3$

The chemical reaction that takes place when nitrogen and hydrogen react is written like this:

N$_2$ + 3H$_2$ → 2NH$_3$

This tells us that one molecule of nitrogen combines with three molecules of hydrogen to give two molecules of ammonia.

Nitrogen and hydrogen react together to make ammonia. This has many uses and is transported around the country in large trucks.

Glossary

acid rain: When certain gases rise into the atmosphere, they dissolve in rainwater, making the rain acidic.

atmosphere: The layer of air around Earth made of nitrogen, oxygen, carbon dioxide, water vapor, and tiny traces of other gases.

atom: The smallest part of an element that still has all the properties of that element.

atomic number: The number of protons in an atom.

bond: The attraction between two atoms that holds the atoms together.

cells: The building blocks from which plants and animals are made.

compound: A substance that is made of atoms of more than one element. The atoms are held together by bonds.

electron: A tiny particle with a negative charge. Electrons are found inside atoms, where they move around the nucleus in layers called electron shells.

inert: Something that does not readily react with other substances.

metal: An element on the left of the periodic table. Metals are good conductors of heat and electricity.

mineral: A compound or element as it is found in its natural form in Earth. Potassium nitrate is a mineral.

molecule: A particle that contains atoms held together by chemical bonds.

nebula: A cloud of gas and dust that exists in the spaces between stars.

neutron: A tiny particle with no electrical charge found in the nucleus of an atom.

nonmetal: An element at the right hand side of the periodic table. Nonmetals are liquids or gases at normal temperatures. They are poor at conducting heat and electricity.

nucleus: The center of an atom. It contains protons and neutrons.

nutrients: The chemicals in food that plants and animals need for healthy growth and development.

ore: A rock that contains a mineral mixed up with other substances.

periodic table: A chart of all the chemical elements laid out in order of their atomic number.

pollution: Dirtying of the air or water by chemicals.

pressure: The amount of force pressing on a given area.

products: The substances formed in a chemical reaction.

proton: A tiny particle with a positive charge. Protons are found inside the nucleus of an atom.

reactants: The substances that react together in a chemical reaction.

smog: A poisonous mixture of fog and chemicals.

Index